S. Carolina Albuquerque de Andrade
Larissa Lima de Sousa

Efficacy of essential oils in controlling Rhizopus stolonifer on grapes

S. Carolina Albuquerque de Andrade
Larissa Lima de Sousa

Efficacy of essential oils in controlling Rhizopus stolonifer on grapes

Use of Rosmarinus officinalis L. Origanum vulgare L. as natural antimicrobials in fruit

ScienciaScripts

Imprint

Any brand names and product names mentioned in this book are subject to trademark, brand or patent protection and are trademarks or registered trademarks of their respective holders. The use of brand names, product names, common names, trade names, product descriptions etc. even without a particular marking in this work is in no way to be construed to mean that such names may be regarded as unrestricted in respect of trademark and brand protection legislation and could thus be used by anyone.

Cover image: www.ingimage.com

This book is a translation from the original published under ISBN 978-613-9-60990-1.

Publisher:
Sciencia Scripts
is a trademark of
Dodo Books Indian Ocean Ltd. and OmniScriptum S.R.L publishing group

120 High Road, East Finchley, London, N2 9ED, United Kingdom
Str. Armeneasca 28/1, office 1, Chisinau MD-2012, Republic of Moldova, Europe
Printed at: see last page
ISBN: 978-620-8-21775-4

SUMMARY

To God, for he is all honor and glory,

To my mother and my sisters, To all my friends, I dedicate this with all my affection.

ACKNOWLEDGMENTS

To God for the gift of life, for always guiding me, lighting my way and even without deserving it, pouring out his blessings on me.

To my mother Izabel Albuquerque, an exemplary woman, a warrior, a fighter, who was by my side in difficult times, advising and teaching me the best way, who never spared any effort for my education, who was always father and mother to her daughters.

My sisters, Eryka Albuquerque, for all their love, care and dedication to me, always by my side. To Isabelle Albuquerque and Michela Albuquerque for their affection and support, and the moments of happiness we shared.

To my aunts and uncles, cousins (especially Alex and Cristiane), brothers and sisters-in-law, nephews and friends Denise, Maeli, Paulo, Ramon, Heyde, Valclemir, Maiara, who, sometimes near and sometimes far, cheered me on and for my professional success.

To my boyfriend Valber Pessoa for all his love, concern and patience, because it was you who I turned to at times when I was desperate and didn't know what to do with the mishaps of academic life; for always encouraging me and showing me that I do have the capacity to fulfill all my professional desires and wishes. Thank you to your family, who have become special people in my life.

To my advisor, Professor Dr. Rita de Càssia Ramos do Egypto Queiroga, for your understanding, help, patience and dedication to this monograph, for the love of the person you are, for your voice that has become my reassurance and for your sweet smile that captivates everyone.

I am immensely grateful to Professor Dr. Evandro Leite de Sousa for the valuable opportunity to be in his research group, for the initiation granted and for sharing with us the example of a researcher that you are.

To my master friend Larissa Souza for the trust and friendship she placed in me, she was a little angel sent from heaven to open up my research paths. Thank you for all the happy moments we spent together doing analysis, traveling, or just chatting away that we loved to have.

To everyone in the Microbiology laboratory, where I was welcomed from day one, and over time they became a family to me. Especially Professor Maria Lúcia for her teaching, dedication and care for her students, Helena who recommended me, Camila who was always available to help me, Carlos Eduardo who was a private tutor for any doubts, Rayanne, Ana Julia, Neusa, Nelson, Nereide, Jossana, Vanessa, Ingrid, Priscila, Dona Salete, you were all part of this growth.

My classmates, it wasn't easy living with only women, but then the men arrived to harmonize. Especially Alenna, Camilla, Giana and Pryscila, who were always with me on this journey, studying for exams, doing assignments and often talking about the anguish they were experiencing and giving me the comfort my heart needed.

To all the teachers on the Nutrition course, for having the pleasure of passing on knowledge with such love and zeal, they have always been my mirror. To all the staff who were always so helpful in assisting us with whatever we needed.

And all those who haven't been mentioned, but who are rooting for me, who have been there in sad and happy times, who have prayed, toasted and rejoiced in my achievements. Thank you from the bottom of my heart!

"Fear not, for I am with thee: be not dismayed, for I am thy God: I will strengthen thee, and help thee, and uphold thee with the right hand of my righteousness." [Isaiah 41:10]

SUMMARY

The Isabel grape is one of the main varieties of *Vitis labrusca*, which has an accelerated reduction in post-harvest quality, due to weight loss due to softening of the tissues, darkening of the skin and the occurrence of off-flavors resulting from the senescence process, as well as sensitivity to dehydration and fungal infection during handling and processing, limiting its marketability. The control of post-harvest fungal diseases is carried out using synthetic fungicides. These chemical compounds have been withdrawn from the market due to possible toxicological risks. Due to the negative impact of these fungicides, the search for alternative methods effective in the control and development of phytopathogens has stimulated research into the use of essential oils for application to foodstuffs. In this context, the present study evaluated the efficacy of applying the essential oils of *Origanum vulgare* L. (OV) and *Rosmarinus officinalis* L. (RO) alone and in combination as inhibitors on the post-harvest fungus *Rizhopus stolonifer* URM 3728 "*in vitro*", the influence on the autochthonous microbiota of Isabel grapes (*Vitis labrusca* L.), and the physical and physical-chemical characteristics of the fruit during storage at room temperature (25 °C for 12 days) and refrigerated temperature (12 °C for 24 days). OV and RO revealed MIC values of 0.25 and 1 µL/mL, respectively. The application of essential oils at different concentrations (CIM RO 1 µL/mL; CIM OV 0.25 µL/mL; 1/2 CIM OV 0.125 µL/mL + 1/2 CIM RO 0.5 µL/mL; 1/4 CIM OV 0.06 µL/mL + 1/4 CIM RO 0.25 µL/mL) inhibited mycelial growth (95.02 - 98.89%) and spore germination (> 63%) of the fungus used in the test, as well as inhibiting the autochthonous mycoflora of grapes stored at room temperature and low temperature. In general, the application of the coating composed of OV and RO at sub-lethal concentrations preserved the quality of the grapes, which were measured by their physical (color, firmness, loss of mass) and physico-chemical (titratable acidity, soluble solids and titratable acidity/soluble solids ratio) attributes. The results obtained in this study show the potential for applying OV and RO at sub-lethal concentrations to control post-harvest fungal pathogens.

Keywords: viticulture, post-harvest, antimicrobial agents.

1 INTRODUCTION

The outstanding growth of agriculture in Brazil is partly due to the large-scale production of fruit, making it the world's second largest producer, especially in terms of tropical fruit, expanding agribusiness and being an important consumer and exporter (SILVEIRA et al., 2005). However, the volume of exports is still considered small due to the high volume of losses, estimated at 10 million tons/year, which comprises an average of 30% of the total fruit produced (DANTAS et al., 2003).

Reducing post-harvest losses in the fruit production chain represents a constant challenge, considering that fruits are organs with a high water and nutrient content and, even after harvesting until senescence, maintain various biological processes in activity, thus presenting a greater predisposition to physiological disorders, mechanical damage and the occurrence of rot (KADER, 2002).

Rot resulting from the activity of pathogens causes serious losses, especially when the fruit is grown in places far from the area of consumption, as the diseases can start in the field and remain latent, only appearing after favorable environmental conditions (DANTAS et al., 2003). Fungal growth depends on a complex interaction between various factors, such as temperature, humidity, water activity, pH, nutritional factors, type of product, time and storage conditions (ANLI; ALKIS, 2010; GARCIA et al., 2011; LASRAM et al., 2010).

The Isabel grape, one of the main varieties of *Vitis labrusca L.,* whose bunches of berries are very clustered and ripen quickly, originated in the south of the United States and was later spread to other regions. It is a hardy variety, less demanding in terms of cultivation and, because it is more tolerant of fungal diseases, it is well adapted to humid climate conditions (DETONI et al., 2005). The grape production chain in Brazil is quite complex, including various sectors such as table grapes (*in natura*), wines and juices, which have been showing a clear growth trend in recent years (RITSCHEL; CAMARGO, 2007).

The grape agro-industrial system is of significant economic importance, due to the numerous jobs in the sectors of inputs, production, processing, distribution and support services (MELLO, 2009). Although grapes are resistant to fungal diseases, the main method used to control fungal pathogens are synthetic fungicides (ROJAS-GRAU; SOLIVA-FORTUNY;

NIARTIN-BELLOSO, 2008), some of which have been withdrawn from the market due to possible toxicity risks to consumers (CALVO et al., 2007).

The increase in public interest and that of health authorities regarding the presence of synthetic fungicide residues in agricultural products, the possible risk of these substances accumulating in the environment (SILVEIRA et al., 2005), as well as concern about the negative consumer perception of synthetic fungicides used for many years to solve the problem of fungal degradation in fruit, and the development of fungal strains, 2005), as well as concern about the negative consumer perception of synthetic fungicides used for many years to solve the problem of fungal decay in fruit, and the development of fungicide-resistant strains, have motivated research into natural compounds as an alternative method for controlling diseases and, consequently, post-harvest losses in fruit (SANTOS et al., 2012; XING et al., 2012).

In this context, studies have focused on plant essential oils and their compounds because they have antimicrobial and antioxidant properties (RASOOLI; REZAEI; ALLAMEH, 2006). The essential oils of oregano (OV) and rosemary (RO) are recognized for not causing any toxic effects to living beings, raising no concerns about the amount used in food (BURT, 2004). The essential oils (OV) and (RO) used alone proved to be promising in terms of antimicrobial activity, but in quantities greater than organoleptically acceptable. Combining different concentrations of essential oils can be a way of improving the antifungal efficiency of the product used and the sensory acceptability of the foods to which they are applied (AZERÊDO et al., 2011).

Considering these aspects, the present study evaluated the efficacy of applying the essential oils of *Origanum vulgare L.* (oregano) and *Rosmarinus officinalis* L. (rosemary) alone and in combination at sublethal concentrations in inhibiting the growth of the post-harvest fungal pathogen *Rizhopus stolonifer* URM 3728 in laboratory conditions and on grapes (*V. labrusca L.*) and its influence on the autochthonous microflora, physical and phyto-chemical characteristics of the fruit during storage at room temperature and refrigerated.

2 OBJECTIVES

2.1 General objective

To evaluate the effect of the isolated and combined application of the essential oils of *Origanum vulgare L.* (oregano) and *Rosmarinus officinalis* L. (rosemary) in inhibiting the growth of post-harvest fungal pathogens and autochthonous microbiota of grapes (*Vitis labrusca*).

2.2 Specific objectives

• To determine the minimum inhibitory concentration (MIC) of oregano and rosemary essential oils against pathogenic fungi of interest in post-harvest;

• Evaluate the antimicrobial potential of oregano and rosemary essential oils against natural fruit microbiota;

• To verify the antimicrobial potential of essential oils against the pathogenic fungus of interest in post-harvest fruit;

• To observe the effects of the application of essential oils on the autochthonous microflora and on the physico-chemical characteristics of the fruit at different temperatures.

3 LITERATURE REVIEW

3.1 Post-harvest quality of grapes

Brazilian viticulture was born with the arrival of the Portuguese colonizers in 1532, where it was just a domestic crop. At the beginning of the 20th century, it became a commercial activity on the initiative of the Italian immigrants who settled in the south of the country. From its beginnings until the 1960s, Brazilian viticulture was restricted to the South and Southeast regions, as they have a temperate climate with an annual growing cycle and low winter temperatures. During this period, grapevines were grown in the semi-arid region of the Vale Submédio Sao Francisco, initiating tropical viticulture in Brazil (LEÂO; SOARES, 2000; PROTAS; CAMARGO; MELLO, 2006).

Tropical viticulture expanded rapidly, solidifying several clusters in various Brazilian regions (South, Southeast, Northeast) (PROTAS; CAMARGO; MELLO, 2006), after there was a greater intensity of planting, modernization of wineries, formation of technological bases focused on the production of table grapes and for the production of wines and juices (GÒES, 2005).

In the Northeast, the main cluster is in the Vale do Submédio Sao Francisco, whose cities are Petrolina, in Pernambuco, and Juazeiro, in Bahia, where they are pioneers in the production of grapes and wines under tropical conditions in Brazil (IBRAVIN, 2012). The Valley is considered the second largest wine producing region in Brazil, accounting for 15% of national production, with eight million liters per year (PEREIRA, 2012). Its irrigated area comprises 120,000 hectares, of which 12,200 hectares are cultivated with vines. Around 95% of the planted area is used to produce grapes for *fresh* consumption (table grapes), both for the domestic market and for export (SOARES; LEÂO, 2009). However, grape juice production in the Middle Valley is still on a small scale (with the South and Southeast regions producing the most) and the product is widely accepted in Brazil, mainly due to its health benefits (SANTOS; FERRAZ, 2006).

Brazilian grape juice production has a promising future. According to a survey carried out by the Brazilian Viticulture Union - UVIBRA (2012), there was an 18.3% increase in production of this product compared to the previous year. This growth has led companies to invest heavily in structure and technology in order to meet demand and improve quality standards. The main cultivars used for making juice are Isabel, Isabel precoce, BRS Rùbea, Concord and Bordò (RITSCHEL; CAMARGO, 2007).

On the international scene, in 2009 Brazilian viticulture ranked 19th in terms of area under grapes and 14th in terms of production. In terms of fruit exports, Brazil was the 17th and 10th largest exporter of grape juice (MELLO, 2011).

According to statistics available on the IBGE website, in 2011 there was a 12.97% increase in grape production in Brazil, with the South region being the country's largest producer, contributing 62% of the total, followed by the Northeast region with 21% (EMBRAPA, 2012).

The main grape crops in Brazil are divided into two groups: one made up of fine table grapes (*Vitis vinifera*), represented mainly by cultivars such as Italia and its mutations (Rubi, Benitaka and Brasil), Red Globe, Red Meire, Patricia and the seedless ones (Centennial Seedless, Superior Seedless or Festival, Thompson Seedless, Perlette, Catalunya and Crimson Seedless); and another for common or rustic table grapes (*Vitis labrusca*), the main representative of which is the Niàgara Rosada cultivar (NACHTIGAL, 2003).

American grapes or common grapes (*Vitis labrusca*) are characterized by high yields and greater resistance to fungal diseases than *Vitis vinifera*, as well as having lower production costs and being sold for less than European grapes. Fine or European grapes (*Vitis vinifera*) have their own characteristics for the production of fine or table wines, and have low resistance to crop diseases, making the use of treatments and pesticide applications greater (NACHTIGAL; CAMARGO; MAIA, 2005).

The grape is a non-climacteric fruit, with low physiological activity, very sensitive to dehydration and fungal infection during post-harvest handling (ARTÉS- HERNÀNDEZ; TOMAS-BARBERAN, 2006). After harvest and throughout storage, the main problems with table grapes are dehydration, degranulation and rot, which can be alleviated by proper and careful handling of the fruit (KUGLE et al., 2002). Some disorders are physiological in nature, almost always caused during refrigerated storage, while others are caused by improper handling of the fruit, leading to infection by microorganisms (OLIVEIRA et al., 2006; CAMARGO et al., 2011).

Post-harvest diseases in grapes come from quiescent or acquired infections. The infectious process in quiescent infections usually begins in the pre-harvest phase, when the fruit is still attached to the plants. However, as soon as they are harvested, they undergo physiological changes, making them more susceptible to attack by pathogens, which are normally at rest (CAVALCANTI et al., 2005). Quiescent infections are often caused by fungi of the genera *Alternaria, Colletotrichum, Lasiodiplodia and Botrytis*. In the case of acquired infections, which occur due to wounds after harvesting, the fungi *Penicillium, Cladosporium, Aspergilus and Rhizopus* stand out, which quickly manifest symptoms of rot (OLIVEIRA et al., 2006).

The attack of these microorganisms causes the occurrence of superficial diseases, destruction of tissues, resulting in a reduction in their quality and shelf life, making them less attractive or unmarketable (MOSS, 2002).

The occurrence of fungal diseases in table grape crops can cause major losses and becomes a limiting factor for viticulture when adequate control measures are not adopted. The susceptibility of the main cultivars planted, the environmental conditions favorable to the development of pathogens, as well as inadequate crop management, mean that grape growing can only be made viable with the massive application of fungicides, increasing production costs, the risk of poisoning workers and contamination of the environment (NAVES; GARRIDO; SÔNEGO, 2006).

Faced with these difficulties, new technological alternatives for production and protection must be evaluated and implemented, but which are feasible, with the aim of improving productivity and fruit quality, minimizing environmental damage and not compromising human health, as well as reducing post-harvest losses.

3.2 Control of post-harvest fungal pathogens

Plant diseases are responsible for considerable losses to crops of economic importance, including post-harvest diseases in vegetables, which are "those that occur after harvest due to the lack of marketing or consumption of the product in a timely manner; in other words, resulting from damage to the crop after it has been harvested, as a result of accumulated damage from the place of production, in addition to that which occurred during transportation, storage, processing and/or marketing of the saleable product" (CHITARRA; CHITARRA, 2005).

Active infections occur when the fruit has started or completed the ripening process, progressing as the environmental conditions favor the growth of the pathogen. In these infections, penetration occurs mainly through wounds caused during harvesting, storage and marketing, although in some cases it can occur even through the intact surface of the fruit (DANTAS et al., 2003).

Diseases caused by fungi are those that occur most frequently during the post-harvest phase of vegetables (CHITARRA; CHITARRA, 2005). Vegetables, due to their low pH, high moisture content and nutrient composition, are very susceptible to the action of fungal pathogens, which in addition to their deteriorating action, can make products unfit for consumption due to the production of mycotoxins, which can have mutagenic, teratogenic or carcinogenic effects on consumers when consumed chronically (FRISVAD; SKOUBOE; SAMSON, 2005; PRAKASH et al., 2010; PRAKASH et al., 2011; REDDY et al., 2011). The fungi that stand out as post-harvest pathogens *are Pénicillium digitatum, P. italicum, Botrytis cinerea, Alternaria alternata, A. solani, Rhizopus stolonifer, Aspergillus flavus and Aspergillus Niger, the* main culprit in the production of ochratoxin

A in grapes and grape derivatives (DAFERERA et al., 2003; TZORTZAKIS; EKONOMAKIS, 2007; PONSONE et al., 2011; LASRAM et al., 2012).

Although it is difficult to determine the total magnitude of post-harvest losses due to microbial spoilage, which vary widely according to the type of vegetable, production area and seasonal aspects, it is known that up to 25% of the world's vegetable production is subject to attack by micro-organisms in industrialized countries, while in developing countries this damage is often higher, reaching up to 50% of production (SPADARO; GULLINO, 2004).

As a way of reducing this damage, physical, chemical and biological methods have been used to control this group of diseases. Chemical control, through the application of synthetic fungicides, remains the main measure for restricting the incidence of post-harvest diseases in vegetables. These compounds can be used alone, combined in mixtures, or applied separately in sequence (ISMAIL; ZHANG, 2004). This method has significant disadvantages, including increased production costs, dangers for handlers, concerns about residues in food, threats to public health and the environment (FERNANDEZ et al., 2001; SOROUR; LARINK, 2001). Furthermore, due to the emergence of new physiological species of phytopathogenic fungi and the development of resistance on the part of these microorganisms, many of the synthetic fungitoxic compounds widely used are gradually becoming less effective (PARANAGAMA et al., 2003).

The widespread use of fungicides in the world is variable, although it is estimated that around 23 million kilos of these compounds are applied to fruit and vegetables every year, and it is generally believed by producers that the production and marketing of perishable plant products would not be possible without their use (RAGSDALE; SISLER, 1994). Carcinogenicity, teratogenicity, high and acute residual toxicity, long degradation period, environmental pollution, influence on the organoleptic characteristics of food, and side effects in humans are the main factors that have restricted the use of chemical fungicides in the control of post-harvest deterioration (UNNIKRISHNAN; NATH, 2002).

Synthetic antifungals belonging to the group of benzimidazoles, aromatic hydrocarbons and sterol biosynthesis inhibitors are frequently used as post-harvest treatments. The application of increasing concentrations of chemical agents in an attempt to overcome the problem of fungal resistance results in the presence of high levels of toxic residues in the products and the environment, as well as an increase in the intensity of their side effects and the final cost of production (FALANDYSZ, 2000; LEELASUPHAKU et al., 2008).

Public concern about these risks has sparked interest in finding safer crop protectants to replace synthetic chemical pesticides. An emerging alternative has been the use of natural protectants with fungitoxic potential, which should have low toxicity to mammals, fewer deleterious effects on the environment and wide public acceptance (LIU; HO, 1999; HAMILTON-KEMP et al., 2000).

The use of so-called natural fungicides appears to be another option to the use of synthetic fungicides in terms of control efficiency (WILSON et al., 1997). They are viable alternatives to the traditional chemical method, mainly because they do not leave toxic residues on the treated products (WILSON; WISNIEWSKI, 1989). Biocontrol, characterized as the use of organisms and/or their derived products or metabolites to prevent plant diseases, is ecologically viable, usually safe, and can provide long-term protection for the crop (SAN-LANG et al., 2002; FERNANDO et al., 2005).

Some agents used in the biocontrol process have been shown to be efficient as an alternative to synthetic fungicides in preventing post-harvest spoilage of vegetables, such as aroma compounds (ARCHBOLD et al., 1999; UTAMA et al., 2002), acetic acid (SHOLBERG et al., 1998), glucosinolates (MARI et al., 2002; MARI et al., 2003), propolis (LIMA et al., 1998), chitosan (CAPDEVILLE et al., 2002; CHOI et al., 2002), plant extracts (MOHAPOTRA et al., 2000; TRIPATHI et al., 2002) and essential oils (CHU et al., 2001).

3.3 Antimicrobial activity of essential oils

The antibacterial and antifungal activity of essential oils has long been recognized, but the food industry has focused more attention on their application as natural antioxidants (PLOOY; REGNIER; COMBRINCK, 2009). Essential oils are recognized for their inhibitory activity against a wide variety of microorganisms, including viruses, fungi, protozoa and bacteria (SOLÓRZANO-SANTOS; MIRANDA-NOVALES, 2012). According to Bhavanani and Ballow (1992), around 60% of essential oils have antifungal action and 30% have antibacterial properties.

Essential oils are obtained from different parts of plants, mainly leaves, through steam or hydrodistillation; and are characterized as a complex mixture of mainly terpenoids, monoterpenes and sesquiterpenes, in addition to a variety of aromatic phenols, oxides, ethers, alcohols, esters, aldehydes and ketones that determine the characteristic aroma and smell of the plant (BATISH et al... 2008), 2008); in addition to providing functions necessary for the survival of the plant itself, they also play an important role in defense against microorganisms (SIQUI et al., 2000).

In general, the action of phenolic compounds on fungi includes cytoplasmic coagulation, disorganization of cell contents, rupture of the plasma membrane, inhibition of fungal enzymes, disturbance of the functionality of genetic material, culminating in inhibition of spore germination and mycelial growth (BURT, 2004; LIU et al., 2007; MENG et al., 2010).

Exposure to terpenes can interfere with the expression of genes encoding virulence factors, such as when considering enterotoxin-producing strains *of S. aureus* (QIU et al., 2011); and with the expression of cytoplasmic and membrane proteins in *Salmonella enterica* (DI PASQUA et al., 2010). The presence of volatile monoterpenes or essential oils provides an important defense strategy for plants, particularly against herbivorous insects and pathogenic fungi (BATISH et al., 2008). Because they are easily extracted, biodegradable and easily catabolized by the environment, essential oils do not persist in soil and water (ISMAN; MACHIAL, 2006), have low or no toxicity to fish, birds and mammals, and therefore form the basis of many applications, including applications in food preservation, pharmaceuticals, alternative and natural medicine (BAKKALI et al., 2008).

Essential oils have attracted scientific interest due to the fact that they are natural products recognized as GRAS (Generally Recognized as Safe), have a broad spectrum of antimicrobial activity, and are effective in controlling pathogenic and spoilage microorganisms of importance in plants (BURT, 2004; OMIDEYGI et al., 2007; GUTIERREZ et al., 2009).

Essential oils have two main characteristics as antimicrobial agents for possible use in food: their natural origin, which means greater safety for consumers and the environment; and they are considered to have a low risk of developing microbial resistance. The second characteristic cited is based on the fact that essential oils are made up of a wide variety of constituents, which apparently have different mechanisms of antimicrobial activity, thus making it more difficult for microorganisms to adapt to their action (DAFERERA et al., 2003).

Lipophilic compounds, such as those present in essential oils, are efficiently absorbed by fungal cells, as the fungal mycelium is largely lipophilic in nature, associated with a large relative surface area compared to its volume (REGNIER et al., 2008). The mechanism underlying the antifungal action of essential oils serves as a switch between the vegetative and reproductive phases of fungal development, since they are cited for negatively impacting sporulation due to the impediment of mycelial development and/or the perception/transduction of different physiological signals involved in the synthesis of molecules for the functioning of the vegetative form, with a view to the development of the reproductive form. As a consequence of the suppression of spore production resulting from exposure to a certain essential oil, the propagation of the pathogen is limited, reducing the release of spores into the environment (TZORTZAKIS; ECONOMAKIS, 2007).

Previous studies have shown that the essential oils of *Origanum vulgare L.* (oregano) and *Rosmarinus officinalis L.* (rosemary) have interesting antimicrobial activities against spoilage and food-related pathogenic microorganisms (OLIVEIRA et al., 2010; SOUZA et al., 2009).

The essential oils *Origanum vulgare L.* (oregano) and *Rosmarinus Officinalis L.* (rosemary) have shown outstanding results in inhibiting different microorganisms, although the results have varied according to the microorganisms tested, the origin of the essential oil and the technique used (DUARTE et al., 2005; SOUZA et al., 2007).

When deciding to apply essential oils to food, it is essential to consider the impact on the sensory characteristics of the products during storage, since the concentration of essential oil required to establish antimicrobial efficacy can result in organoleptic characteristics that are unpleasant for consumption (HSIEH et al., 2001; GUTIERREZ et al., 2009). Considering these aspects, the use of essential oils to inhibit microbial growth in foods has been the subject of research and improvement, since these natural antimicrobials can also, when applied in excessive doses, exceed the limits of consumer acceptance, with implications for both the taste and smell of the products (SOUZA et al., 2007).

Therefore, research in this area should be focused on optimizing essential oils and their applications to obtain sufficiently effective antimicrobial activity at low concentrations so as not to adversely affect the organoleptic properties and acceptability of food (GUTIERREZ, 2009).

3.4 Combined application of essential oils and phytoconstituents

Previous studies have found that essential oils have more than one mechanism of action (BURT, 2004; MOREIRA, 2005), suggesting that the combined application of essential oils could result in high efficacy against the action of pathogenic microorganisms in food (GUTIERREZ et al., 2008). Thus, the combined application of antimicrobial compounds must have an additive resultant effect compared to the individual effect, i.e. synergism occurs when the effect of the combined substances is greater than the sum of the individual effects (BURT, 2004).

Gutierrez, Barry-Ryan and Bourke (2008) and Lambert et al. (2001) carried out studies combining oregano and thyme essential oils, and found that when the oils are used in combination, the concentrations required to achieve the desired antimicrobial efficacy may be lower than when used alone, thus minimizing the impact on the sensory attributes of food. Apostolidis et al. (2008) found that although there was a higher content of phytochemicals in oregano essential oil when applied alone, the efficacy against *L. monocytoneges* was more intense when the essential oil was applied in combination with *cranberry* extract.

Lv et al. (2011) tested four combinations of essential oils obtained from various spices (oregano + basil, basil + bergamot, oregano + perilla and oregano + bergamot), and all the combinations tested caused cellular destruction of the micro-organisms (*Escherichia coli, Staphylococcus aureus,*

Bacillus subtilis and Saccharomyces cerevisae). Other combinations of essential oils with other antimicrobials have also been successful in inhibiting various micro-organisms, such as the study carried out by Santos et al. (2012) using oregano essential oil with chitosan against *A. niger* and *Rhizopus stolonifer*; as well as Rohani et al. (2011) verified the inhibition of L. *monocytogenes* with garlic essential oil with nisin, and Gutierrez et al. (2008) combined oregano and thyme essential oil, as well as oregano, marjoram, thyme and sage in the inhibition *of Bacillus cereus, Pseudomonas aeruginosa, Escherichia coli O157:H7 and L. monocytogenes.*

Although the antifungal properties of volatile compounds from higher plants have been reported, little attention has been paid to the fungitoxic effect of these substances when applied in combination. This information is desirable, since the fungitoxic potential of most fungicides has been reported to be enhanced when these substances are applied in combination (LEVY et al, 1986;. GULLINO; GARIBALDI, 1987; MIGHELI et al, 1988; PANDEY; DUBEY, 1997).

A study by Fu et al. (2007) investigated the possible interaction between clove and rosemary essential oils at sub-inhibitory concentrations and found that the antimicrobial effect was influenced not only by the ratio of oils applied, but was also, and above all, microorganism-dependent. The researchers found an additive antimicrobial effect against *S. epidermidis, S. auresus, B. subtilis, E. coli, Proteus vulgaris and P. aeruginosa; a* synergistic effect against *Candida albicans* and an antagonistic effect against *A. niger.*

Thus, according to Gutierrez (2009), studies in this field should emphasize the optimization of combinations of essential oils and their components in order to obtain effective antimicrobial activity in the post-harvest protection of fruits and vegetables, even when applied in low concentrations, so as not to adversely affect the organoleptic properties and acceptability of the food.

4 MATERIALS AND METHODS

4.1 Materials

The essential oils of *O. vulgare* (oregano) and *R. officinalis* (rosemary) were supplied by Aromalândia Ind. Com. Ltda. (Minas Gerais, Brazil), and their quality parameters were described in a technical report issued by the supplier.

Isabel table grapes (*Vitis labrusca*) at a stage of commercial ripeness were obtained from EMPASA (Empresa Paraibana de Abastecimentos e Serviços, Joao Pessoa, Brazil) in the morning and selected without signs of mechanical damage or deterioration, standardized in small bunches with grapes of homogeneous size, color and shape.

The strain of the filamentous fungal post-harvest pathogen *Rhizopus stolonifer* URM 3728 was obtained from the Culture Collection of the Mycology Department of the Biological Sciences Center of the Federal University of Pernambuco (UFPE). For the antifungal activity tests, replicates of stock cultures subcultured in test tubes containing Sabouraud agar (Himedia, India) and incubated at 25-28 °C for seven days for sufficient sporulation were used. The fungal spores were harvested by adding sterile saline solution (NaCl 0.85 g/100 mL) to the fungal growth medium, and the suspension obtained was filtered through a triple layer of sterile gauze to retain the fragments of hyphae. The number of spores present in the suspension was then determined by counting in a Newbauer chamber. The concentration of spores obtained was adjusted with sterile saline solution to provide a fungal inoculum of approximately 106 spores/mL (RASOOLI; ABYANETH, 2004; RASOOLI; OWLIA, 2005).

4.2 Determination of the Minimum Inhibitory Concentration of essential oils

The Minimum Inhibitory Concentration of the essential oils was determined using the broth dilution technique. Initially, 1 mL of the fungal suspension was inoculated into 4 mL of Sabouraud broth supplemented with a concentration adjusted to 10 mL. Then 5 mL of the solution with different concentrations of essential oils (8%, 4%, 2%, 1%, 0.5%, 0.25%, 0.125%) was added. Each system was incubated at 25-28 °C for 72 hours. At the end of the incubation period, the lowest concentration (highest dilution) of the oil that showed no visible fungal growth was considered the Minimum Inhibitory Concentration (SHARMA; TRIPATHI, 2007).

Based on the results obtained in the CIM determination tests, the different concentrations of the essential oils of *O. vulgare* L. (OV) and *R. officinalis* (RO) were chosen alone and *in combination* (CIM RO *1μL/mL;* CIM OV *0μL/mL;* CIM OV *0μL/mL*). *officinalis* (RO) alone and in combination

(CIM RO 1 µL/mL; CIM OV 0.25 µL/mL; 1/2 CIM OV 0.125 µL/mL + 1/2 CIM RO 0.5 µL/mL; 1/4 CIM OV 0.06 µL/mL + 1/4 CIM RO 0.25 µL/mL) which were used in the mycelial mass and spore germination quantification tests.

4.3 Influence of fungal mycelial mass

The inhibition of fungal mycelial mass (dry weight) was determined using the technique of poisoning the growth substrate (broth dilution technique). 10 mL of the fungal suspension (approximately 10^6 spores/mL) was inoculated into an Erlenmeyer flask containing 90 mL of Sabouraud broth with the different concentrations of essential oils added, and then incubated at 25-28 °C. After 3, 5 and 7 days of exposure, the mycelial mass obtained was dehydrated in an oven at 60 °C for six hours, followed by maintenance at 40 °C for a further 18 hours (RASOOLI AND ABYANEH, 2004; RASOOLI REZAEI; ALLAMEH, 2006). In the control experiment, the suspension of the filamentous fungus was inoculated into Sabouraud broth without the addition of the essential oil concentrations. By comparing the weight of the dry mycelial mass of the suspension treated with the essential oils and the control test, the percentage of inhibition of the fungal mycelial mass was obtained at the different time intervals.

4.4 Interference with fungal spore germination

Aliquots of 0.1 mL of each solution containing different concentrations of essential oils were mixed with 0.1 mL of fungal spore suspension (10^6 spores/mL) obtained from ten-day-old cultures grown on Sabouraud agar at 25-28°C. Next, 0.1 mL of the system was placed in the center of sterile glass slides, which were then incubated in a humid chamber at 25-28 °C for 24 hours. After this period, each slide was fixed with lactophenol blue cotton dye and observed under an optical microscope for spore germination. Approximately 200 spores were counted from each slide. The effectiveness of the essential oils in inhibiting the germination of fungal spores was observed by comparing the number of spores germinated in the solutions with the added concentrations of the oils with the number of spores germinated in the control experiment, where the solution of these agents was replaced with Sabouraud broth (FENG; ZENG, 2007).

Following the results obtained in the mycelial mass and spore germination tests, the different concentrations of OV and RO were chosen (CIM RO 1 µl ./ml.; CIM OV 0.25 µL/mL; ^{1}/2 CIM OV 0.125 µL/mL +^{1}/2 CIM RO 0.5 µL/mL; ^{1}/4 CIM OV 0.06 µL/mL +^{1}/4 CIM RO 0.25 µL/mL) which were used in the trials on Isabel grapes.

4.5 Effect on fruit decay control

Initially, the grapes were immersed in a solution of sodium hypochlorite (1% v/v) for 15 minutes, washed with drinking water and left to dry in the air for two hours. The fruits were then immersed for one minute in the broths containing the different concentrations of essential oils and dried on nylon mesh to drain off any excess liquid. After this procedure, the fruits were placed in polyethylene containers with lids, so that one group was stored at room temperature (25 °C), while the other group was stored under cooling conditions (12 °C). At the same time, a control experiment was carried out in which the essential oil solutions were replaced with sterile distilled water. Each treatment included 40 fruits, and at different storage time intervals (room temperature 0, 2, 4, 6, 8, 10 and 12 days; cooling 0, 3, 6, 9, 12, 15, 18 and 21 days) the fruits were examined for the presence of visible fungal infection, with the results expressed as the percentage of infected fruits at the different intervals analyzed (FENG; ZENG, 2007; LIU et al., 2007).

4.6 Physico-chemical analysis

At the same intervals of observation of the presence of fungal infection, the fruit was evaluated for weight loss and physical aspects (color and firmness) and physico-chemical aspects (soluble solids, titratable acidity and SS/ AT ratio).

4.6.1 Soluble solids (SS)

The soluble solids content was determined using a digital refractometer model HI 96801 (Hanna Instruments, Sao Paulo, Brazil), with the results expressed in °Brix (MENG et al., 2008).

4.6.2 Titratable acidity (TA)

Titratable acidity was determined using phenolphthalein as an indicator using 0.1N NaOH, with the results expressed in mmol H+/100 g of fruit (MENG et al., 2008).

4.6.3 SS/AT ratio

The SS/ATT ratio was based on the ratio of soluble solids content to titratable acidity content (MENG et al., 2008).

4.6.4 Loss of mass

The fruit was placed on trays so that they remained separated in equal proportions (control and treatment) and the difference in final weight was monitored during the experiment. The results were expressed as a percentage of the initial mass of the product (MENG et al., 2008).

4.6.5 Coloring

Peel color was measured at three different equatorial positions on the fruit, using the CIELab System (L*a*b*); Hue angle (h*ab) and chroma (C*ab) on a Minolta Model CR-300 colorimeter (Osaka, Japan), according to the International Commission on Illumination (CIE, 1986).

4.6.6 Firmness

Firmness was determined using a 3 mm diameter probe (1/8) coupled to a TA-XT2 Texture Analyzer (Stable Micro Systems, Haslemere, England), with the results expressed in N/mm (CHIEN et al., 2007).

4.7 Statistical analysis

All the analyses were carried out in triplicate (one grape from the apical portion, one from the middle portion and one from the bottom of each bunch) in three replicates, with the results expressed as the average of the data obtained in each replicate. Statistical analysis was carried out using descriptive statistics (mean and standard deviation) and inferential tests (ANOVA followed by Tukey's test) to determine statistically significant differences (P < 0.05) between the treatments applied using Sigma Stat software. 2.03.

5 RESULTS

The essential oils of *Origanum vulgare L.* (OV) and *Rosmarinus officinalis* L. (RO) showed Minimum Inhibitory Concentration values of 0.25 μL/mL and 1 μL/mL respectively, against the fungal strain of *R. stolonifer* as shown in **Table 1.**

Table 1. MIC of *Origanum vulgare L.* (OV) and *Rosmarinus officinalis* L. (RO) essential oils against *Rizhopus stolonifer 3728.*

	O. vulgare (μL/mL)	*R. officinalis* (μL/mL)
Cepa	**CIM**	**CIM**
Rhizopus stolonifer	0,25	1

The CIF values for the combined application of OV and RO were 0.5 μL/mL for all the fungi tested, indicating the synergism of the antimicrobial compounds, meaning that when the compounds are applied in combination, even in small quantities, they show greater activity than when applied alone. Spore growth was observed when the essential oils of OV and RO were applied alone at sub-lethal concentrations (1/2 and 1/4). This was not the case when the oils were used in combination, as fungal growth was inhibited even at sub-inhibitory concentrations such as ¼ OV MIC + ¼ RO MIC.

Using the results of the synergy tests, different concentrations of OV and RO were chosen (0.25 μL /mL OV; 1 μL /mL RO; 0.12 μL /mL OV + 0.5 μL /mL RO; 0.06 μL/mL OV + 0.25 μL /mL RO), which were used separately or in combination in tests to inhibit mycelial growth, spore germination and application to grapes. After seven days of incubation, OV and RO strongly inhibited fungal mycelial growth with values of up to 95, 80 %, as can be seen in **Table 2.**

Table 2. Percentage of inhibition of the growth of the fungal mycelial mass of *Rhizopus Stolonifer* 3728 in liquid medium over 7 days of exposure to different combinations of essential oils of *O. vulgare* L. (OV) and *R. officinalis* L. (RO).

Rhizopus stolonifei' 3728	% Inhibition		
		Days	
Treatment	3	5	7
CIM OV	97,88	97,67	95,64
CIM RO	96,89	97,61	95,18
¹/2 CIM OV +¹ /2 CIM RO	96,21	98,57	95,02
1/4 TOP + 1/4 TOP	97,80	98,89	95,80

Results expressed as percentage inhibition of mycelial mass growth compared to the control experiment (0 µL/mL).

MIC O: Minimum Inhibitory Concentration of *Origanum vulgare* L. essential oil;

MIC A: Minimum Inhibitory Concentration of the essential oil of *R. officinalis* L.

The application of OV and RO in the different combinations tested resulted in inhibition of the germination of *R. stolonifer* spores (**Table 3**). The inhibition of the germination of the spores of the test strain in all the concentrations tested was greater than 60% in relation to the number of spores germinated in the control experiment, inhibiting 100% of the spores in the combinations (1/2 OV + ½ RO; ¼ OV + 1/4 RO). These results reinforce the evidence that the combined application of essential oils, or their components, results in increased effectiveness against the action of pathogenic microorganisms.

Table 3. Percentage of inhibition of germination of *Rhizopus stolonifer* 3728 spores exposed to different combinations of *O. vulgare* L. (OV) and *R. officinalis* L. (RO) essential oils for 24 hours.

Treatment	*Rhizopus stolonifer* 3728
CIM OV	95,00
CIM RO	63,00
¹/2 CIM OV +¹ /2 CIM RO	100,00
¹/4 CIM OV +¹ /4 CIM RO	100,00

Results expressed as percentage inhibition of spore germination compared to the control experiment (0 μL/mL).

MIC O: Minimum Inhibitory Concentration of *Origanum vulgare* L. essential oil;

MIC A: Minimum Inhibitory Concentration of the essential oil of *R. officinalis* L.

With regard to the control of fruit deterioration, the application of OV and RO essential oils, at the different concentrations tested, resulted in the inhibition of fungal growth at room and refrigerated temperatures. In the control experiment at room temperature, fungal growth was observed from the sixth day, while in the other concentrations of the essential oils, it was observed after the eighth day. At refrigerated temperature, the appearance of visible fungal infection on the grapes occurred on the twelfth day for the control experiment and on the fifteenth day for the concentrations of OV and RO tested.

The physical and physico-chemical changes in grapes not coated or coated with OV and RO alone or in combination were evaluated during storage at room temperature (25 °C) and refrigeration temperatures (12 °C) (**Tables 4 and 5**). At most of the storage intervals evaluated, no differences (P > 0.05) were observed in the firmness of grapes coated and uncoated with the essential oils and stored at refrigerated temperature. However, over the course of the intervals, the grapes tended to lose their firmness naturally, due to dehydration and degranulation. The titratable acidity values of grapes coated with OV and RO alone or in combination and stored at room temperature or cold did not differ (P > 0.05) from the values found for the control experiment during all the storage intervals evaluated. The grapes that were and were not treated with essential oils and stored at room temperature showed higher titratable acidity values during the first few days of storage and, as the days went by, these values decreased.

Grapes submitted and not submitted to OV and RO treatments alone or in combination did not differ (P > 0.05) in terms of soluble solids content at any of the storage intervals evaluated at ambient or refrigerated temperatures, and the same behavior was found for weight loss (weight loss of around 25% at the end of the storage period) (**Table 6 and 7**).

For the initial storage intervals, both at room temperature and refrigerated, the SS/AT ratio for the grapes coated with OV or RO were equal ($P > 0.05$) to the values found for the control experiment. Grapes submitted and not submitted to treatments with the essential oils increased ($P < 0.05$) the SS/AT ratio values with longer storage intervals at both storage temperatures tested.

The color of the grapes coated and uncoated with the essential oils, alone or in combination, was predominantly purple at the different intervals at both storage temperatures tested, since the a* and b* values (positive values close to zero) did not differ significantly ($P > 0.05$) between the fruits. Furthermore, in all treatments, there was a decrease ($P < 0.05$) in the h* values and a concomitant increase in the C* value, revealing a color change from blue to purple in the fruit and a maintenance of its brightness, which was identified from the L* reading.

Table 4: Average values of the physical and physico-chemical quality characteristics of Isabel grapes coated with *O. vulgare* L. and *R. officinalis* L. essential oils alone and in different combinations, and stored at room temperature (25 °C) for 12 days.

Treatments	Exhibition days						
	0	2	4	6	8	10	12
Firmness (N/mm)							
CTR	48,39 (±1.5)	45,42 (±1.9)	5,70 (±0.7)	5,01 (±0.7)	5,02 (±0.8)[b]	5,21 (±1.1)[b]	4,89 (±0.3)[b]
OV ¼ CIM + RO ¼ CIM	53,67 (±1.3)	48,40 (±1.5)	5,20 (±0.7)	4,11 (±0.5)	4,78 (±0.6)[b]	5,45 (±1.1)[b]	4,87 (±1.5)[b]
OV ½ CIM + RO ½ CIM	49,87 (±1.5)	43,64 (±1.8)	5,40 (±0.8)	4,45 (±0.9)	3,92 (±0.7)[a]	3,80 (±1.3)[a]	3,81 (±0.8)[a]
OV CIM	50,76 (±1.0)	49,64 (±1.5)	4,50 (±1.2)	4,71 (±0.4)	4,57 (±0.8)[b]	4,71 (±0.8)[b]	4,69 (±0.7)[b]
RO CIM	49,78 (±1.5)	42,31 (±1.1)	4,92 (±0.9)	4,90 (±0.7)	5,31 (±0.8)[b]	5,74 (±0.8)[b]	4,59 (±1.2)[b]
Titratable acidity (mmol H+/100 g of fruit)							
CTR	51,33 (±1.9)	38,60 (±3.8)	34,62 (±1.1)	33,75 (±2.3)	33,75 (±2.2)	33,73 (±1.1)	36,60 (±0.0)
OV ¼ CIM + RO ¼ CIM	48,00 (±1.1)	36,67 (±1.9)	39,65 (±1.1)	39,40 (±1.1)	38,51 (±1.1)	39,05 (±0.0)	32,02 (±0.0)
OV ½ CIM + RO ½ CIM	40,90 (±1.1)	34,50 (±1.1)	33,30 (±1.9)	36,25 (±1.9)	36,40 (±2.2)	31,34 (±0.0)	33,81 (±2.1)
OV CIM	46,8 (±0.0)	34,30 (±2.2)	32,90 (±1.9)	33,50 (±3.3)	32,50 (±2.2)	33,70 (±1.1)	32,90 (±1.9)
RO CIM	41,41(±3.1)	38,60 (±1.8)	34,60 (±1.1)	38,45 (±2.6)	33,70 (±2.9)	38,60 (±0.0)	38,75 (±0.0)
Soluble Solids (°Brix)							
CTR	1,40 (±0.1)	1,23 (±0.1)	1,33 (±0.1)	1,80 (±0.1)	1,80 (±0.1)	1,60 (±0.2)	1,73 (±0.1)
OV ¼ CIM + RO ¼ CIM	1,30 (±0.1)	1,50 (±0.0)	1,50 (±0.0)	1,50 (±0.2)	1,70 (±0.1)	1,60 (±0.1)	1,90 (±0.2)
OV ½ CIM + RO ½ CIM	1,30 (±0.1)	1,40 (±0.1)	1,40 (±01)	1,33 (±0.1)	1,63 (±0.2)	1,70 (±0.1)	1,80 (±0.0)
OV CIM	1,40 (±0.1)	1,40 (±0.1)	1,40 (±0.1)	1,30 (±0.2)	1,40 (±0.1)	1,53 (±0.1)	1,50 (±0.3)
RO CIM	1,40 (±0.0)	1,30 (±0.1)	1,30 (±0.1)	1,40 (±0.3)	1,90 (±0.1)	1,40 (±0.3)	1,33 (±0.3)

CTR: control (OV 0 µL/mL + RO 0 µL/mL); OV ¼ CIM: 0.06 L/mL + RO CIM: 0.25 µL/mL; OV ½ CIM: 0.125 µL/mL + RO ½ CIM: 0.5 µL/Ml; OV CIM: 0.25 µL/mL; RO CIM: 1 µL/mL.[a-c] For each test, different superscript lower-case letters in the same column denote a significant difference (P <0.05) between the values obtained according to the Tukey test.

Table 5 Average values of the physical and physico-chemical quality characteristics of Isabel grapes coated with essential oils of *O. vulgare* L. and *R. officinalis* L. alone and in different combinations, and stored at low temperature (12 °C) for 24 days.

Treatments	Exhibition days								
	0	3	6	9	12	15	18	21	24
	Firmness (N/ mm)								
CTR	48,37 (±1.4)	47,40 (±1.3)	5,01 (±0.8)	4,33(±0.8)	4,28 (±0.5)	4,25 (±0.7)	3,64 (±0.6)	3,83 (±0.4)	2,86 (±0.5)[b]
OV ¼ CIM + RO ¼ CIM	53,49(±1.7)	43,75 (±1.1)	4,11 (±0.5)	4,95 (0.8)	4,89 (±0.1)	4,03 (±0.5)	3,65 (±0.3)	3,50(±0.6)	3,54 (±0.9)[b]
OV ½ CIM + RO ½ CIM	54,37(±1.7)	52,90 (±1.0)	4,45 (±0.9)	4,60 (0.9)	4,12 (±0.6)	4,01 (±0.2)	4,10 (±1.0)	3,69(±0.7)	3,55 (±0.5)[b]
OV CIM	51,24(±1.0)	48,64 (±1.5)	4,71 (±0.4)	4,34 (0.4)	4,76 (±1.6)	4,03 (±0.5)	3,65 (±0.9)	3,25 (±0.8)	2,02 (±0.5)[a]
RO CIM	53,39 (±1.5)	52,44 (±0.4)	4,90 (±0.7)	4,98 (1.0)	4,86(±0.4)	3,80 (±0.2)	3,86 (±0.9)	3,30(±1.0)	3,26 (±1.5)[b]
	Titratable Acidity (mmol g of H+/100 fruit)								
CTR	52,33 (±1.9)	46,93 (±1.9)	47,01 (±1.1)[b]	46,80 (±1.1)[b]	47,70 (±1.0)[b]	46,40 (±1.2)[b]	47,50 (±1,2)[c]	36.65(±1.9)[a,b]	37.95 (±1.9)[a,b]
OV ¼ CIM + RO ¼ CIM	48,00 (±1.1)	47,10 (±2.2)	42,00 (±1.0)[a]	43,90 (±2.2)[b]	44,30 (±1.8)[b]	42,30 (±1.1)[b]	47,64 (±1,1)[c]	41,20 (±1.9)[b]	41,10 (±1.9)[b]
OV ½ CIM + RO ½ CIM	40,90 (±1.1)	40,00 (±1.1)	47,81 (±1.1)[b]	43,14 (±1.9)[b]	46,53 (±2.2)[b]	39,43 (±1.4)[b]	38,74 (±2,2)[b]	31,90(±1.1)[a]	33,30 (±2.3)[a]
OV CIM	46,80 (±0.0)	46,90 (±1.9)	47,20 (±2.2)[b]	47,75 (±2.2)[b]	48,20 (±1.2)[b]	46,43 (±16)[b]	47,00 (±0,0)[c]	37,70 (±2.2)[b]	38,30 (±1.1)[b]
RO CIM	43,10 (±2.2)	42,70 (±1.1)	41,00 (±1.9)[a]	38,15 (±1.9)[a]	33,65 (±1.1)[a]	34,84 (±1.9)[a]	33,94 (±2,2)[a]	33.05(±2.4)[a,b]	32,93 (±1.9)[a]
	Soluble Solids (°Brix)								
CTR	1,40 (±0.1)	1,50 (±0.1)	1,50 (±0.2)	1,30 (±0.1)	1,50 (±0.1)	1,31 (±0.2)	1,43 (±0.1)	1,70 (±0.2)	1,80 (±0.2)
OV ¼ CIM + RO ¼ CIM	1,30 (±0.1)	1,33 (±0.2)	1,43 (±0.1)	1,43 (±0.1)	1,60 (±0.1)	1,42 (±0.1)	1,60 (±0.2)	1,70 (±0.1)	1,53 (±0.1)
OV ½ CIM + RO ½ CIM	1,0 (±0.1)	1,40 (±0.1)	1,60 (±0.1)	1,60 (±0.0)	1,50 (±0.1)	1,48 (±0.2)	1,43 (±0.1)	1,80 (±0.1)	1,40 (±0.2)
OV CIM	1,40 (±0.1)	1,30 (±0.1)	1,43 (±0.1)	1,50 (±0.1)	1,70 (±0.1)	1,51 (±0.3)	1,50 (±0.1)	1,60 (±0.2)	1,43 (±0.2)
RO CIM	1,40 (±0.0)	1,20 (±0.2)	1,40 (±0.2)	1,50 (±0.2)	1,20 (±0.2)	1,53 (±0.1)	1,60 (±0.1)	1,33 (±0.3)	1,70 (±0.1)

CTR: control (OV 0 µL/mL + RO 0 µL/mL); OV ¼ CIM: 0.06 µL/mL + RO CIM: 0.25 µL/mL; OV ½ CIM: 0.125 µL/mL + RO ½ CIM: 0.5 µL/Ml; OV CIM: 0.25 µL/mL; RO CIM: 1 µL/mL. [a-c] For each test, different superscript lower-case letters in the same column denote a significant difference (P <0.05) between the values obtained according to the Tukey test.

Table 6. Mass loss values of Isabel grapes (*Vitis Labrusca*) described by percentage at refrigeration temperature (25 °C) during 12 days of storage.

Time (Days)	Loss of mass (%)				
	Control	25	50	CIM O	CIM A
0	100	100	100	100	100
2	94,07	94,96	94,73	94,91	93,79
4	89,99	81,85	90,75	90,47	88,87
6	86,40	78,89	86,45	86,12	92,73
8	83,39	75,93	82,74	82,24	73,55
10	79,73	73,17	71,44	72,46	71,03
12	74,86	71,78	70,37	70,11	70,23

Table 7. Mass loss values of Isabel grapes (*Vitis Labrusca*) described by percentage at refrigeration temperature (12 °C) during 24 days of storage.

Time (Days)	Loss of mass (%)				
	Control	25	50	CIM O	CIM A
0	100	100	100	100	100
3	95,78	97,25	96,38	96,95	97,16
6	92,78	94,79	93,77	94,77	95,47
9	90,47	92,85	91,75	92,41	93,28
12	89,40	91,47	90,45	90,54	91,31
15	87,88	89,87	88,79	88,68	89,48
18	86,64	88,89	87,54	85,82	88,25
24	83,78	85,82	84,18	83,76	85,16

6 DISCUSSION

Rhizopus is one of the fungal genera reported by several authors as being the predominant microbiota in grapes harvested in Argentina, Brazil and France (SAGE et al., 2002; BAU et al., 2005; BELLÌ et al., 2006), however *Rhizopus, Fusarium* and *Trichoderma* occur in specific stages and areas (FRESJ et al., 2007). Castro et al. (1999) reported in their study that the species *P. expasum, A. alternata and R. stolonifer* occur widely on grapes produced in the Vale do Submédio Sao Francisco region (PE) and in the Jales region (SP).

The combined application of OV and RO at different sub-inhibitory concentrations caused inhibition of fungal growth in liquid media, as well as causing changes in the morphology (and germination) of *R. stolonifer* spores and mycelia. Inhibition of spore germination was greater when the fungus tested was exposed to mixtures of OV and RO, which is an important finding because the spore is known to be an important structure for the survival and proliferation of fungi (RABEA et al, 2009). Therefore, the application of compounds that strongly inhibit spore germination or cause the destruction of these structures are interesting candidates for the control of spore populations that spread over various substrates or environments (PLASCENCIA-JATOMEA et al, 2003; SHARMA AND TRIPATHI, 2008).

In a previous study, the combination of chitosan and OV in sub-lethal quantities caused significant morphological alterations in the spores (shriveling, disruption, loss of genetic material and deepening of furrows) and mycelium (wear, wrinkles and loss of cytoplasmic material) of *R. stolonifer* and *A. niger* (SANTOS et al., 2012). The authors suggested that the mechanism by which chitosan and OV inhibited spore germination was the result of their interaction with the spore cell wall.

Essential oils have shown remarkable power to inhibit the growth of post-harvest fungal pathogens (XING et al., 2012; OMIDBEYGI et al., 2007). Carmo et al. (2008) and Mitchell et al. (2010) reported inhibition of *Aspergillus* species (*A. parasiticus, A. terreus, A. flavus, A. ochraceus and A. parasiticus*) by *O. vulgare* essential oil (0.6 - 80 µL/mL) over 14 days of exposure at 25 °C.

The application of OV and RO at sublethal concentrations (1/2 CIM + 1/2 CIM; ¼ CIM + 1/4 CIM) resulted in greater inhibition of *R. stolonifer* (at germination) on grapes in the autochthonous fungal microbiota than the effects caused by the application of both essential oils alone, which suggests a synergistic antifungal effect of OV and RO. The synergy of OV and RO against post-harvest pathogenic fungi has not been reported previously. Azerêdo et al. (2011) found a synergistic effect of the combined application of OV and RO against the bacteria associated with minimally processed vegetables, *Listeria monocytogenes, Yersinia enterocolitica and Aeromonas hydrophylla*. According to these authors, the increase in antimicrobial activity caused by the mixture of OV and RO could be partially explained by considering the different compounds found in each essential oil.

In general, essential oils that have strong antimicrobial properties contain high amounts of carvacrol and thymol. Azerêdo et al. (2011) found significant amounts (carvacrol, 66.9%), and also suggest that hydroxyl groups increase the antimicrobial properties of essential oils and, therefore, 1,8- cineol, which is the most prevalent compound in the RO tested in this study (cineol 32.2%), may also contribute to establishing the rapid and constant antimicrobial effect achieved with the combination of essential oils. In a later study, Sousa et al. (2012) also found a synergistic effect of carvacrol and 1,8-cineole against *L. monocytogenes, Y. enterocolitica, A. hydrophylla and P. fluorescens.* Minor components found in essential oils may also play a critical role in establishing the synergistic antimicrobial effect of OV and RO (GUTIERREZ et al, 2009). Therefore, the hydrocarbons (such as α-pinene, camphene, myrcene, α-terpinene and p-cymene), which are known to possess weak antimicrobial properties, found in RO (AZERÊDO et al., 2011), in quantities greater than 1% appear to swell microbial cells, therefore allowing carvacrol to be easily transported into the cell (GUTIERREZ et al., 2008).

In a study carried out by Santos et al. (2012), chitosan and OV were found to have an inhibitory effect using sub-lethal amounts of chitosan and OV against *R. stolonifer* and *Aspergillus niger* in artificially infected table grapes as well as the autochthonous fungal microbiota of grapes (*Vitis Labrusca*) stored at ambient and cold temperatures.

Bearing in mind that the physical and physico-chemical parameters of fruit are essential quality parameters that influence their shelf life and consumer acceptance, this study evaluated the effects of OV and RO alone and in combination in sub-lethal quantities, which maintained and/or improved these attributes in table grapes at room temperature and in cold storage. In agreement with the results obtained in this study, other researchers have reported that the application of essential oils alone or in combination preserved or improved the physical and physico-chemical quality of fruit (SANCHEZ-GONZALEZ et al, 2011; HASSANI et al, 2012). AZÊREDO et al, 2011;. SOUSA et al, 2012;. SANTOS et al, 2012)

7 CONCLUSION

According to the results obtained in this study, it was possible to reveal that oregano essential oil (*Origanum vulgare L.*) and rosemary essential oil (*Rosmarinus officinalis L.*), when applied in combination at different sub-inhibitory concentrations, have a remarkable ability to inhibit the post-harvest pathogenic fungi *R. stoionifer* and the indigenous fungal microbiota in grapes (*Vitis Labrusca*) stored at cooling and ambient temperatures. It was also observed that the combined application of the tested compounds did not negatively affect the physical and physico-chemical characteristics of the grapes, and that some of their characteristics improved during the storage period. These findings indicate the potential of the combined application of OV and RO at sub-inhibitory concentrations to control the growth of fungal pathogens in fruit, and could emerge as an alternative to the synthetic antifungals currently applied in order to reduce post-harvest losses due to the action of fungal pathogens.

REFERENCES

ANLI, E.; ALKIS, I. M. Ochratoxin A in wines. **Food Reviews International**, v.25, n.1, p. 214-232, 2009.

APOSTOLIDIS, E.; KWON, Y. I.; SHETTY, K. Inhibition of *Listeria monocytogenes* by oregano, cranberry and sodium lactate combination in broth and cooked ground beef systems and likely mode of action through proline metabolism. **International Journal of Food Microbiology**, v. 128, p. 317-24, 2008.

ARCHBOLD, D. D.; HAMILTON-KEMP, T. R.; CLEMENTS, A. M.; COLLINS RANDY, W. Fumigating "Crimson Seedless" table grapes with (E)-2-hexenal reduces mold during long-term postharvest storage. **Hort Science**, v.34, p. 705-707, 1999.

ARTÉS-HERNÀNDEZ, F.; TOMAS-BARBERAN, F. A. Modified atmosphere packaging preserves quality of SO2-free "Superior seedless" table grapes. **Postharvest Biology and Technology**, v. 39, p. 146-154, 2006.

AZÊREDO, G.A.; STAMFORD,T. L. M.; NUNES, P. C.; GOMES NETO, N. J.; OLIVEIRA, M. E. G.; SOUZA, E. L. Combined application of essential oils from *Origanum vulgare L.* and *Rosmarinus officinalis* L. to inhibit bacteria and autochthonous microflora associated with minimally processed vegetables. **Food Research International**, v. 44, p. 1541-1548, 2011.

BAKKALI, F.; AVERBECK, S.; AVERBECK, D.; IDAOMAR, M. Biological effects of essential oils - a review. **Food Chemistry Toxicology**, v. 46, p. 446-475, 2008.

BATISH, D. R.; SINGH, H. P.; KOHLI, R. K.; KAUR, S. Eucalyptus essential oil as a natural pesticide. **Forest Ecology and Management**, v. 256, p. 2166-2174, 2008.

BAU, M. et al. Ochratoxigenic species from Spanish wine grapes. **International Journal of Food Microbiology**, v. 98, n. 2, p. 125-130, 2005.

BHAVANANI, S. M.; BALLOW, C. H. New agents for Gram-positive bacteria. **Current Opinion in Microbiology**, v. 13, p. 528-534, 1992.

BELLÌ, N et al. Mycobiota and ochratoxin A producing fungi from Spanish wine grapes. **International Journal of Food Microbiology**, v. 111, n.1, p. 46-52, 2006.

BURT, S. Essential oils: their antibacterial properties and potential applications in foods - a review. **International Journal of Food Microbiology**, v. 94, p. 223-253, 2004.

CALVO, J.; CALVENTE, J.; ORELLANO, M. E.; BENUZZI, D.; TOSETTI, M. I. S. Biological control of postharvest spoilage caused by *Penicillium expansum* and *Botrytis cinerea* in apple by using the bacterium *Rahnella aquatilis*. **International Journal of Food Microbiology**, v.113, p.251-257, 2007.

CAMARGO, R. B.; PEIXOTO, A. R.; TERAO, D.; ONO, E. O.; CAVALCANTI, L. S. Fungi causing post-harvest rot in apirenic grapes in the agricultural center of Juazeiro-BA and Petrolina-PE. **Revista Caatinga,** v. 24, p. 15-19, 2011.

CAPDEVILLE, G.; DE WILSON, C. L.; BEER, S. V.; AIST, J. R. Alternative disease control agents induce resistance to blue mold in harvested Red Delicious apple fruit. **Phytopathology**, v. 92, p. 900-908, 2002.

CARMO, E. S., LIMA, E. O., SOUZA, E. L. The potential of *Origanum vulgare L.* (Lamiaceae) essential oil in inhibit the growth of some food-related *Aspergillus species.***Brazilian Journal of Food Microbiology**, v. 39, p. 362-367, 2008.

CASTRO, J. V. de et al. Use of packaging for post-harvest preservation of grapes. **Revista Brasileira de Engenharia Agricola e Ambiental**, Campina Grande, v. 3, n. 1, p. 35-40, 1999.

CHIEN, P.; SHEU, F.; YANG, F. Effects of edible coating on quality and shelf life of sliced mango fruit. **Journal of Food Engineering**, v. 78, p. 225-229, 2007.

CHITARRA, M. I. F.; CHITARRA, A. B. **Post-harvest of fruit and vegetables**. Editora UFLA, Lavras, p. 783, 2005.

CHU, C.L.; LIU, W.T.; ZHOU, T. Fumigation of sweet cherries with thymol and acetic acid to reduce post harvest brown rot and blue mold rot. **Fruits**, v.56, p.123-130, 2001.

CAVALCANTI, L. S.; DI PIERO, R. M.; CIA, P.; PASCHOLATI S. F.; RESENDE, M. L. V.; ROMEIRO, R.S. **Induction of resistance in plants to pathogens and insects.**Piracicaba: FEALQ, p. 263, 2005.

CHOI, W. Y.; PARK, H. J.; AHN, D. J.; LEE, J.; LEE, C. Y. Wettability of chitosan coating solution on Fiji apple skin. **Journal of Food Science**, v. 67, p. 2668 - 2672, 2002.

CIE - Commission Internationale de l'Éclairage. **Colorimetry**. Vienna: CIE publication, 1986.

DAFERERA, D. J.; ZIOGAS, B. N.; POLISSIOU, M. G. The effectiveness of plant essential oils on the growth of *Botrytis cinerea, Fusarium sp.* and *Clavibacter michiganensis subsp. michiganensis.***Crop Protection**, v. 22, p. 39-44, 2003.

DANTAS, S. A. F.; OLIVEIRA, S. M. A.; MICHEREFF, S. J.; NASCIMENTO, L. C.; GURGEL, L.M.S.; PESSOA, W.R.L.S. Postharvest fungal diseases in papaya and oranges sold at the Recife Central Supply Station. **Fitopatologia Brasileira**, v. 28, p. 528-533, 2003.

DETONI, A. M.; CLEMENTE, E.; BRAGA, G. C.; HERZOG, N. F. M. Niàgara Rosada grapes grown in the organic system and stored at different temperatures.

Ciência Tecnologia Alimentos, v. 25, p. 546-552, 2005.

DI PASQUA, R.; MAMONE, G.; FERRANTI, P.; ERCOLINI, D.; MAURIELLO, G. Changes in the proteome of *Salmonella enterica* serovar Thompson as stress adaptation to sublethal concentrations of thymol. **Proteomics,** v.10, p. 1040-9, 2010.

DUARTE, M. C. T.; FIGUEIRA, G. M.; SARTORATTO, A.; GARCIA, V. L.; DELARMELINA, C. Anti-Candida activity of Brazilian medicinal plants. **Journal of Ethnopharmacology**, v. 97, p. 305-311, 2005.

EMBRAPA -**Brazilian** Vitiviniculture: Panorama 2011 - ISSN 1808-680, Comunicado115 Técnico - 2012.

FALANDYSZ, J. Residues of hexachlorobenzene in Baltic fish and estimation of daily intake of this compound and pentachlorobenzene with fish and fishery products in Poland. **Polish Journal of Environmental Studies,** v. 9, p. 377-383, 2000.

FENG, W.; ZENG, X. Essential oils to control *Alteranaria alternate* in vitro and in vivo **Food Control**, v. 18, p. 1126-1130, 2007.

FERNANDEZ, M.; PICO, Y.; MANES, J. Pesticides residues in Orange from Valencia (Spain). **Food Additives and Contaminants**, v. 18, p. 614-624, 2001.

FERNANDO, C. J.; SKOUBOE, P.; & SAMSON, A. R. Taxonomic comparison of three different groups of aflatoxin producers and new efficient producer of aflatoxin B1, sterigmatocystine and 3-OmethylsterigmatocystineAspergillus *rumbelli sp.* nov.

Systematic and Applied Microbiology, v. 28, p. 442-453, 2005.

FREDJ,S. M. B.; CHEBIL, S.; MILIK, A. Occurrence of pathogenic fungal species in Tunisian vineyards. **International Journal of Food Microbiology**, v. 113, n. 1, p. 245250, 2007.

FU Y.J.; ZU Y.G.; CHEN L.Y.; SHI X.H.G.; WANG Z.; SUN S. Efferth T. Antimicrobial Activity of clove and rosemary essential oils alone and in combination. **Phytother Research**, v. 21, p. 989-999, 2007.

GARCIA, D. et al. Is intranspecific variability of growth and mycotoxin production dependent on environmental conditions?: a study with Aspergillus carbonarius isolates.

International Journal of Food Microbiology, v. 144, n. 3, p. 432-439, 2011.

GÓES, F. J. **Development and optimization of the fermentation process for the production of white wine from Itàlia grapes**. 158p. Dissertation (master's degree in Chemical Engineering) - Federal University of Sao Carlos, Sao Carlos, 2005.

GULLINO, M.L.; GARIBALDI, A. Control of *Botrytis cinerea* resistant to benzimidazoles and dicarboximides with mixtures of different fungicides.

Mededelingen Van de Faculteit Landbouwwenschappen Rijkuniversitiet Gent, v. 52, p.895-900, 1987.

GUTIERREZ, J., BARRY-RYAN, C., BOURKE, P. The antimicrobial efficacy of plant essential oil combinations and interactions with food ingredients. **International Journal of Food Microbiology**, v. 124, p. 91-97, 2008.

GUTIERREZ, J.; BARRY-RYAN, C.; BOURKE, P. Antimicrobial activity of plant essential oils using food model media: efficacy, synergistic potential and interactions with food components. **Food Microbiology**, v. 26, p. 142-150, 2009.

HAMILTON-KEMP, T.R.; McCRAKEN Jr., C.T.; LOUGHRIM, J. H.; ANDERSON, R. A.; HIDELBRAN, D. F. Effect of some natural volatiles on the pathogenic fungi *Alternaria alternata* and *Botrytis cinerea*. **Journal of Chemical Ecology**, v.18, p.10831091, 2000.

HASSANI, A., FATHI, Z., GHOSTA, Y., ABDOLLAHI, A., MESHKATALSADAT, M. H., MARANDI, R. J. Evaluation of plant essential oils for control of postharvest brown and gray mold rots on apricot. **Journal of Food Safety**, v. 32, p. 94-101, 2012.

HSIEH, P. C.; MAU, J. L.; HUANG, S. H. Antimicrobial effect of various combinations of plant extracts. **Food Microbiology**, v. 22, p. 35-43, 2001.

BRAZILIAN WINE INSTITUTE. **Producing regions**. Available at: <http: www.ibravin.org.br/regioesprodutoras.hp> Accessed: 29 Mar. 2013

ISMAIL, M.; ZHANG, J. Post-harvest citrus diseases and their control outlook. **Pest Management.** v. 15, p. 29-35, 2004.

ISMAN, M.B., MACHIAL, C.M.. Pesticides based on plant essential oils: from traditional practice to commercialization. In: Rai, M., Carpinella, M.C. (Eds.), Naturally Occurring Bioactive Compounds. **Advances in Phytomedicine**, vol. 3, p. 29-44 , 2006.

KADER, A. (ed.) **Postharvest Technology of Horticultural Crops**. 3 ed. Riverside: UC Regents, 2002. 535p.

KUGLE, R. A. et al. **Physiology and post-harvest management of temperate climate fruits**. Campinas: Livraria e Editora Rural, p. 214, 2002.

LASRAM, S. et al. Evolution of ochratoxin A content during red and rose vinification. **Journal of the Science of Food Microbiology**, v. 114, n. 3, p. 376-379, 2007.

LASRAM, S. et al. Evolution of ochratoxin A content during red and vinification. **Journal of the Science of Food and Agriculture**, London, v. 88, n.10, p. 1696 - 1703, Aug.2008.

LEÂO, P. C. S.; SOARES, J. M. A **viticultura no semiàrido brasileiro**. Petrolina: EMBRAPA Semiàrido, p. 44, 2000.

LEELASUPHAKUL, W.; HEMMANEEA, P.; CHUENCHITT , S. Growth inhibitory properties of *Bacillus subtilis* strains and their metabolites against the green mold pathogen (*Penicillium digitatum Sacc.*) of citrus fruit. **Postharvest Biology and Technology**, v. 48, p. 113-121, 2008.

LEVY, Y.; BENDERLY, M.; COHEN, Y.; GISI, U.; BASAND, D. The joint action of fungicides in mixtures: comparison of two methods for synergy calculation. **EPPO Bull**, v. 16, p. 651-657, 1986.

LIMA, G.; DE CURTIS, F.; CASTORIA, R.; PACIFICA, S.; DE CICCO, V. Additives and natural products against post harvest pathogens compatibility with antagonistic yeasts. In: Plant Pathology and Sustainable Agriculture. **Proceedings of the Sixth SIPaV Annual Meeting**, p. 17-18, 1998.

LIU, Z. L.; HO, S. H. Bioactivity of the essential oil extracted from Evodia rutaecarpa Hook f. et Thomas against the grain storage insects,Stiphilus zeamais Motsch. and

Tribolium castaneum (Herbst.). **Journal of Stored Products Research**, v. 35, p. 317328, 1999.

LIU, J.; TIAN, S.; MENG, X.; XU, Y. Effects of chitosan on control of postharvest diseases and physiological responses of tomato fruit. **Postharvest Biology and Technology,** v. 44, p. 300-306, 2007.

LV, F.; LIANG, H.; YUAN, Q.; LI, C. In vitro antimicrobial effects and mechanism of action of selected plant essential oil combinations against four food-related microorganisms. **Food Research International**, v. 44, p. 3057-64, 2011.

MARI, M.; LEONI, O.; LORI, R.; CEMBALI, T. Antifungal vapour-phase activity of allyl isothiocyanate against Penicillium expansum on pears. **Plant Pathology**, v. 51, p. 231-236, 2002.

MARI, M.; BERTOII, P.; PRATEIIA, G.C. Non-conventional methods for the control of post harvest pear diseases. **Journal of Applied Microbiology**, v. 94, p. 761-766, 2003.

MELLO, C. E. C. The history of wine in Brazil. **Adega magazine**, ed.61, 2009.

MELLO, L. M. R. **Brazil's role in the world wine market**: panorama 2011: technical communiqué.

Brasilia: EMBRAPA, 2011. Available at:

<http://www.agencia.cnptia.embrapa.br/Repositorio>. Accessed on: 12 Jul 2013.

MENG, X.; LI, B.; LIU, J.; TIAN, S. Physiological responses and quality attributes of table grape fruit to chitosan preharvest spray and postharvest coating during storage.

Food Chemistry, v. 106, p. 501-508, 2008.

MENG, X.; YANG, L.; KENNEDY, J. F.; TIAN, S. Effects of chitosan and oligochitosan on growth of two fungal pathogens and physiological properties in pear fruit. **Carbohydrate Polymers**, v. 81, p. 70-75, 2010.

MIGHELI, Q.; ALOI, C.; GULLINO, M.L. Evaluation of the in vitro activity of diethoferrocarb phenyl carbamate against some pathogens sensitive or resistant to benzimidazole. **Difesa Piante**, v. 11, p. 3-12, 1988.

MITCHELL, T. C., STAMFORD, T.L.M., SOUZA, E.L., LIMA, E.O., CARMO, E. S. *Origanum vulgare* L. essential oil as inhibitor of potentially toxigenic Aspergilli.

Food Science and Technology, v. 30, p. 755-760, 2010.

MOHAPOTRA, N. P.; PATI, S. P.; RAY, R. C. In vitro inhibition of Botryodiplodia theobromae (Pat.) causing Java black rot in sweet potato by phenolic compounds.

Annals of Plant Protection Science, v. 8, p. 106-109, 2000.

MOREIRA, M.; PONCE, A.; DEL VALLE, C.; ROURA, S. Inhibitory parameters of essential oils to reduce a foodborne pathogen. **LWT-Food Science and Technology**, v. 38, p. 565-570, 2005.

MOSS, M. O. Mycotoxins review. 1. Aspergillus and Penicillium. **Mycologist**, v. 16, p. 116-119, 2002.

NACHTIGAL, J. C. **Technological advances in table grape production**. X Brazilian Congress of Viticulture and Oenology, Bento Gonçalves, 2003.

NACHTIGAL, J. C.; CAMARGO, U. A.; MAIA, J. D. G. Effect of growth regulators on apyrenic grapes, cv. BRS Clara. **Revista Brasileira de Fruticultura**, v. 27, n. 2, p. 304-307, 2005.

NAVES, R.L.; GARRIDO, L.R.; SÔNEGO, O.R. **Control of fungal diseases in table grapes in the northwest region of the State of São Paulo**. Embrapa Uva e Vinho, Technical Circular 68, Bento Gonçalves, RS December, 2006.

OLIVEIRA, S. M. A. et al. **Post-harvest pathology: tropical fruit, vegetables and ornamentals.** Brasilia: Embrapa Informaçao Tecnològica, p. 855, 2006.

OLIVEIRA, C. E. V.; STAMFORD, T. L. M.; GOMES NETO, N. J.; SOUZA, E. L. Inhibition of *Staphylococcus aureus* in broth and meat broth using synergies of phenolics and organic acids. **International Journal of Food Microbiology**, v. 137, p. 308-311, 2010.

OMIDBEYGI, M.; BARZEGAR, M.; HAMIDI, Z.; NAGHDIBADI, H. Antifungal activity of thyme, summer savory and cloves essential oil against *Aspergillus flavus* in liquid medium and tomato paste. **Food Control**, v.18, n.12, p.1518-1523, 2007.

PANDEY, V.N.; DUBEY, N.K. Synergistic activity of extracted plant oils against Pythium aphanidermatum and P. debaryanum.**Tropical Agriculture**, v. 74, p. 164-167, 1997.

PARANAGAMA, P. A.; ABEYSEKERA, K. H. T.; ABEYWICKRAMA, K.; NUGALIYADD, L. Fungicidal and anti-aflatoxigenic effects of the essential oil of *Cymbopogon citratus* (D.C.) Stapf. (lemongrass) against *Aspergillus flavus* Link. isolated from stored rice. **Letters in Applied Microbiology**, v. 37, p. 86-90, 2003.

PEREIRA, G. E. **Technical notes**. Curitiba: CODEVASF, 2006. Available at: < http: www.vinhovasf.com.br/site/arquivos/NotasTecnicas.pdf> Accessed on: 24 Apr. 2013.

PLASCENCIA-JATOMEA, M., VINIEGRA, G., OLAYO, R., CASTILLO-ORTEGA, M. M., SHIRAI, K. Effect of chitosan and temperature on spore germination of *Aspergillus niger*. **Macromolecular Bioscience**, v. 3, p. 582-586, 2003.

PLOOY, W.; REGNIER, T.; COMBRINCK, S. Essential oil amended coatings as alternatives to synthetic fungicides in citrus postharvest management. **Postharvest Biology and Technology**, v. 53, p. 117-122, 2009.

PONSONE, M. L. et al. Biocontrol as a stratregy to reduce the impact of ochratoxin A and *Aspergillus* sction Nigri in grapes. **International Journal of Food Microbiology**, v. 151, n. 2, p. 70 - 77, 2012.

PRAKASH, B.; SHUKLA, R.; SINGH, P.; KUMAR, A.; MISHRA, P. K.; & DUBEY, N. K. Efficacy of chemically characterized Piper betle L. essential oil against fungal and aflatoxin contamination of some edible commodities and its antioxidant activity. **International Journal of Food Microbiology**, v. 142, p. 114-119, 2010.

PRAKASH, B., SHUKLA, R., SINGH, P., MISHRA, P. K., DUBEY, N. K., &KHARWAR, R. N. Efficacy of chemically characterized *Ocimum gratissimum L.* essential oil as an antioxidant and a safe plant based antimicrobial against fungal and aflatoxin B1 contamination of spices. **Food Research International**, v. 44, p. 385-390, 2011.

PROTAS, J. F. S.; CAMARGO, U. A.; MELLO, L.M. R. **A viticultura brasileira**: realidade e perspectiva.Brasilia: EMBRAPA, 2001.

QIU, J.; ZHANG, X.; LUO, M.; LI, H.; DONG, J.; WANG, J.; LENG, B.; WANG, X.; FENG, H.; REN, W.; DENG, X. Subinhibitory Concentrations of Perilla Oil Affect the expression of Secreted Virulence Factor Genes in *Staphylococcus aureus*. **PLoS ONE**, v. 6,p. e16160, 2011.

RABEA, I. E., BADAWAY, M. E. I., STEURBAUT, W., & Stevens, C. V. In vitro assessment of N-(benzyl) chitosan derivate against some plant pathogenic bacteria and fungi. **European Polymer Journal**, v. 45, p. 237-245, 2009.

RAGSDALE, N. N.; SISLER, H. D. Social and political implications of managing plant diseases with decreased availability of fungicides in United States. **Annual Review of Phytopathology**, v. 32, p. 545-557, 1994.

RASOOLI, I.; ABYANEH, M. R. Inhibitory effect of Thyme oils on growth and aflatoxin production by *Aspergillus parasiticus*. **Food Control**, v. 15, p. 79-483, 2004.

RASOOLI, I.; OWLIA, P. Chemoprevention by thyme oils of *Aspergillus parasiticus* growth and aflatoxin production. **Phytochemistry**, v. 66, p. 2851-2856, 2005.

RASOOLI, I.; REZAEI, M. B.; ALLAMEH, A. Growth inhibition and morphological alterations of *Aspergillus niger* by essential oil from Thymus eriocalyx and Thymus x- porlock. **Food Control**, v.

17, p. 359-364, 2006.

REGNIER, T.; PLOOW, W. du; COMBRINCK, S.; BOTHA, B.; Fungitocity of Lippia scaberrima essential oil and selected terpenoid components on two mango postharvest spoilage pathogens. **Postharvest Biology and Technology**, v. 48, p. 254-258, 2008.

RITSCHEL, P.; CAMARGO, U. A. **The grape improvement program and the juice segment**. Bento Gonçalves: EMBRAPA Uva e Vinho, 2007. Available at: <http://www.cnpuv.embrapa.br/publica/artigos/melhoramento_suco.pdf>. Accessed on: July 10, 2013.

ROHANI, S. M. R.; MORADI, M.; MEHDIZADEH, T.; SAEI-DEHKORDI, S.S.; GRIFFITHS M. W. The effect of nisin and garlic (*Allium sativum L.*) essential oil separately and in combination on the growth of *Listeria monocytogenes*. **Lwt - Food Science and Technology**, v. 44, p. 2260-2265, 2011.

ROJAS-GRAU, M. A.; SOLIVA-FORTUNY, R.; NIARTIN-BELLOSO, O. Effect of natural antibrowning agents on color and related enzymes in fresh-cut Fuji apples a san alternative to use of ascorbic acid. **Journal of Food Science**, v. 73, p. 267-272, 2008.

SAGE, L. et al. Fungal flora and ochratoxin A production in grapes and musts from France. **Journal of Agricultural and Food Chemistry**, v. 50, n. 5, p. 1306 - 1311, 2002.

SANCHEZ-GONZALES, L., CHAFER, M., CHIRALT, A., GONZALEZ

MARTiNEZ, C. Physical properties of edible chitosan films containing bergamot essential oil and their inhibitory action on *Penicillium italicum*. **Carbohydrate Polymers**, v. 82, p. 277-283, 2010a.

SAN-LANG, W.; SHIN, I.-L.; WANG, C.-H. ; TSENG, K.-C.; CHANG, W.-T.; TWU, Y.-K.; RO, J.-J.; WANG, C.-L. Production of antifungal compounds from chitin by Bacillus subtilis. **Enzyme and Microbial Technology**, v. 31, p. 321-328, 2002.

SANTOS, E. O.; FERRAZ, Z. M. L. The good performance of Bahian fruit growing: agrosynthesis. **Bahia Agricola**, Salvador, v. 7, p. 3-10, 2006.

SANTOS, N. S. T.; AGUIAR, A. J. A. A.; OLIVEIRA, C. E. V.; SALES, C. V.;

SILVA, S. M.; SILVA, R. S.; STAMFORD, T. C. M.; SOUZA, E. L. Efficacy of the application of a coating composed of chitosan and *Origanum vulgare L.* essential oil to control *Rhizopus stolonifer* and *Aspergillus niger* in grapes (*Vitis labrusca L.*).**Food Microbiology**, v. 32, p. 345 - 353, 2012.

SHARMA, N.; TRIPATHI, A. Effects of *Citrus sinensis (L.)* Isbeck epicarp essential oil on growth and morphogenesis of *Aspergillus niger*.**Microbiological Research**, v. 163, p. 337-344, 2007.

SHARMA, N., TRIPATHI, A. Effects of *Citrus sinensis* (L.) Osbeck epicarp essential oil on growth and morphogenesis of *Aspergillus niger* (L.) Van Tieghem.

Microbiological Research, v. 163, p. 337-344, 2008.

SHOLBERG, P. L. Fumigation of fruit with short chain organic acids to reduce the potential of post harvest decay. **Plant Disease**, v. 82, p. 689-693, 1998.

SILVEIRA, N. S. S.; MICHEREFF, S. J.; SILVA, I. L. S. S.; OLIVEIRA, S. M. A. Postharvest fungal diseases in tropical fruits: pathogenesis and control. **Revista Caatinga**, v. 18, n. 4, p. 283-299, 2005.

SIQUI, A. C.; SAMPAIO, A. L. F.; SOUSA, M. C.; HENRIQUES, M. G. M. O.; RAMOS, M. F. S. Essential oils - anti-inflammatory potential. **Biotecnologia, Ciência e Desenvolvimento**, v. 16, p. 38-43, 2000.

SOARES, J. M.; LEÂO, P.C.S. **A vitivinicultura no semiàrido brasileiro.** Petrolina: EMBRAPA Informaçao Tecnològica, p. 756, 2009.

SOLÓRZANO-SANTOS; MIRANDA-NOVALES. Essential oils from aromatic herbs as antimicrobial agents. **Current Opinion in Biotechnology**, v. 23, p. 136-141, 2012.

SOROUR, J.; LARINK, O. Toxic effects of benomyl on the ultrastructure during spermatogenesis of the earthworms Eisenia fetida. **Ecotoxicology and Environmental Safety,** v. 50, p. 180-188, 2001.

SOUZA, E. L.; STAMFORD, T. L. M.; LIMA, E. O.; TRAJANO, V. N. Effectiveness of *Origanum vulgare L.* essential oil to inhibit the growth of food spoiling yeasts. **Food Control**, v. 18, p. 409-413, 2007.

SOUZA, E. L.; BARROS, J. C. B.; CONCEIÇÂO, M. L.; GOMES NETO, N. J.; & COSTA, A. C.

V. Combined application of *Origanum vulgare L.* essential oil and acetic acid for controlling the growth of *Staphylococcus aureus* in foods.**Brazilian Journal of Microbiology**, v. 40, p. 387-393, 2009.

SOUSA, J. P., AZERÊDO, G. A., TORRES, R. A., CONCEIÇÂO, M. L., VASCONCELOS, M. A. S., SOUZA, E. L.Synergies of carvacrol and 1,8-cineole to inhibit bacteria associated with minimally processed vegetables. **International Journal of Food Microbiology**, v. 158, p. 9-13, 2012.

SPADARO, D.; GULLINO, M. L. State of the art and future prospects of the biological control of postharvest fruit diseases. **International Journal of Food Microbiology**, v. 91, p. 185-194, 2004.

TRIPATHI, P.; DUBEY, N. K.; PANDEY, V. B. Kaempferol: the antifungal principle of Acacia nilotica Linn. **Journal of the Indian Botanic Society**, v. 81, p. 51-54, 2002.

TZORTZAKIS, N. G.; ECONOMAKIS, C. D. Antifungal activity of lemongrass (*Cymbopogon citratus L.)* essential oil against key postharvest pathogens. Innovative **Food Science and Emerging Technology**, v. 8, p. 253-258, 2007.

BRAZILIAN VITICULTURE Union. Sales of wines and derivatives produced in RS - 2006 à 2011 - domestic and foreign markets, in liters.

Available at: http://www.uvibra.com.br/. Accessed on: 29 Apr. 2013.

UNNIRISHNAM, V.; NATH, B. S. Hazardous chemicals in food. **indian Journal of Dairy Bioscience,** v. 11, p. 155-158, 2002.

UTAMA, I. M. S.; WILLS, R. B. H.; BEN-YE-HOSHUA, S.; KUEK, C. in vitro efficacy of plant volatiles for inhibiting the growth of fruit and vegetable decay microorganisms. **Journal of Agricultural and Food Chemistry**, v. 50, p. 6371-6377, 2002.

WiLSON, C.L.; WiSNiEWSKi, M.E. Biological control of postharvest diseases of fruits and vegetables: an emerging technology. **Annual Reviews Phytopathology**, v. 27, p. 425-441, 1989.

WILSON, C.L.; SOLAR, J.M.; GHAOUTH, A.E.; WINIEWSKI, M.E. Rapid evaluation of plant extracts and essential oils for antifungal activity against *Botrytis cinerea*. **Plant Disease**, v. 81, p. 204-210, 1997.

XING, Y., XU, Q., LI, X., CHE, Z., & YUN, J. Antifungal activities of clove oil against *Rhizopus nigricans, Aspergillus flavus* and *Penicillium citrinum* in vitro and in wounded fruit test. **Journal of Food Safety**, v. 32, p. 84-93, 2012.